Gabriel Coffy Rubin
Daniel Callili

Growing Formosa papaya in a tropical climate and sandy soil

Gabriel Coffy Rubin
Daniel Callili

Growing Formosa papaya in a tropical climate and sandy soil

Report on activities carried out in the field

ScienciaScripts

This book is a translation from the original published under ISBN 978-613-9-61892-7.

Publisher:
Sciencia Scripts
is a trademark of
Dodo Books Indian Ocean Ltd. and OmniScriptum S.R.L publishing group

120 High Road, East Finchley, London, N2 9ED, United Kingdom
Str. Armeneasca 28/1, office 1, Chisinau MD-2012, Republic of Moldova, Europe
Printed at: see last page
ISBN: 978-620-7-70642-6

Summary

ACKNOWLEDGMENTS

To my family, Adalberto Rossatto Rubin, Màrcia Coffy Rubin and Ane Coffy Rubin, for all their love, encouragement and affection, which helped me achieve my goals.

To my advisor Prof.[a] Dr. Sarita Leonel, for awakening in me a curiosity for fruit growing and for the knowledge shared.

To the agronomist and owner Thiago Bresinski Lage, for the opportunity to work as an intern at his company.

To my internship supervisor, agricultural engineer Marcos Souza Leal, for his friendship, patience, learning opportunities and all the support I needed during my internship.

To all the employees of the Canta Galo farm, for the moments of conviviality, which contributed directly or indirectly to my training as a professional.

To Rodrigo Moreira, for the friendship created during my internship.

To my friends in Botucatu/SP, for their conviviality and true friendship during the course.

INTRODUCTION

The aim of this report is to explain the activities carried out during the period of the compulsory supervised internship, which was extremely important for my professional growth. I was able to experience the day-to-day production of papaya, which, according to the activity plan, focused on all the stages of growing the crop. I acquired knowledge from planting to the post-harvest stages, activities that enabled me to make decisions and produce quality fruit. The internship was supervised by agronomist Marcos Souza Leal, a graduate of the Minas Gerais School of Administrative Studies (FEAD), and lasted 480 hours.

1. THE COMPANY

The internship was carried out on the Canta Galo farm, located on the BR-349 highway between Bom Jesus da Lapa and Santa Maria da Vitória, KM 38, in the municipality of Serra do Ramalho, in the state of Bahia, which belongs to the company Frutsi Brasil located in the municipality of Janaùba-MG. The company was founded on December 12, 2007 and its main activity is banana cultivation, followed by papaya and mango production.

Fazenda Canta Galo is located on the banks of the Corrente River, a tributary of the Sâo Francisco River, at an altitude of 483 meters, with an average annual temperature of 25.4°C, an average annual rainfall of 833 mm, with September being the hottest month and June the coldest in the region, and its production is restricted to papaya, both for export and for sale on the domestic market. It has a total area of 4560 hectares, where 600 hectares comprise the cultivated area, irrigated using a central pivot system, these 600 hectares are subdivided into six pivots of 100 hectares each. The farm has headquarters with an administrative office, garage, machine workshop, industrial kitchen, cafeteria, three fruit packing sheds, a warehouse for storing chemical products and a storage shed for fertilizers (Fig. 1). The Canta Galo farm employs an average of two hundred permanent employees per month and has three buses of its own to transport employees to the farm every day (Fig. 2).

Figure 1 - Administrative office and workshop (A). Warehouse for storing chemical products and

machine refueling area (B), Canta Galo farm, BA, 2018.

Figure 2: Farm bus fleet, Canta Galo farm, BA, 2018.

2. GROWING PAPAYA

The papaya tree (Carica papaya L.) has its center of origin in South America, where other genera of the Caricaceae family also originate (COSTA et al., 2013). The papaya tree is characterized as a typically tropical plant (BADILLO, 1971). The species Carica papaya L. is a perennial herbaceous plant, in which there are fast-growing hermaphroditic and dioecious individuals that reach up to eight meters in height. However, when it comes to commercial populations, the majority are hermaphrodite plants that can have a lifespan of up to three years (BADILLO, 1993). Brazil is ranked as the second largest producer of papaya in the world, with 35,500 hectares accounting for 15.7% of world production. Most of this production is concentrated in Bahia and Espirito Santo (FRANCO, 2013).

Gyno-andromonoecious (hermaphroditic) papaya trees can be classified according to the size and origin of the fruit into two distinct groups: the Solo group and the Formosa group. The Solo group is represented by strains, while the commercial genotypes of the Formosa group correspond to F1 hybrids. The F1 hybrid called 'Tainung 1' has the characteristic of producing relatively tall plants, with elongated fruit on the hermaphrodite plants and round-elongated fruit on the female plants. The weight of the fruit varies from 900 to 1,100 grams, it tastes great, has good durability, is resistant to transportation and produces an average of 180 t/ha/year (COSTA et al., 2013). The 'Tainung 1' hybrid described above is the one grown on the Canta Galo farm, where the internship took place.

To achieve good plant growth and quality fruit production, the papaya tree requires certain climatic conditions, such as: planting the crop in sunny, warm areas, low altitudes and well-distributed rainfall. The soil must be well-drained and fertile (CARDOSO; PACHECO, 2013).

In regions with high sunshine, the papaya tree grows regularly and produces good quality fruit, as it is a tropical climate fruit (LYRA, 2007). Temperatures between 22°C and 28°C are the ideal average for the crop, with 25°C being the optimum average annual temperature for development (OLIVEIRA et al., 1994).

The crop needs well-distributed annual rainfall of between 1800 and 2000 mm, which is considered ideal for the good development of the crop and, if the producing region does not offer these characteristics, it is necessary to supplement the water with irrigation.

(SANCHES, 2012). Considering morphological aspects such as broad leaves, tall herbaceous stems and heavy fruit production, the crop can be considered vulnerable to the action of strong winds. For the best development of the papaya tree, the soil best suited to its cultivation is sandy-clay loam, with a pH of 5.5 to 6.7 (SIQUEIRA; BOTREL, 1986).

Some observations should be made when deciding whether to plant papaya in a particular region, such as: local climatic conditions conducive to the development and production of the fruit, adequate and well-distributed rainfall during the year or the availability of water for irrigation. Soils with compacted or acidified layers should be corrected before the crop is planted. Finally, the production system must be such as to enable high yields and fruit of excellent quality (CARDOSO; PACHECO, 2013).

3. ACTIVITIES CARRIED OUT

3.1. Soil preparation for planting

Soil preparation for planting papaya is carried out in an east-west direction, in order to make better use of sunlight. It begins with clearing the area, removing native vegetation that would compete with the papaya tree for water, light and nutrients. After the area has been cleared of weeds, subsoiling is carried out, on dry soil, in order to better decompress the soil and facilitate root development. Once the soil has been decompressed, the planting line furrower is used to open up the planting furrows, improving the initial development of the pivoting roots of the papaya tree (Fig. 3). The furrowing operation is carried out at a distance of 3.50 meters between planting lines.

Figure 3. Area after furrowing, Canta Galo farm, BA, 2018.

Once the furrows for planting the seedlings have been opened, basic fertilization is carried out, which consists of adding mineral fertilizers to the soil in order to improve the initial development of the crop. 375 grams of simple superphosphate fertilizer are used per linear meter, consisting of essential elements for plant development, such as phosphorus, calcium and sulphur. The simple superphosphate is distributed in the planting furrow via an implement attached to the tractor (Fig. 4).

Figure 4 - Distribution of simple superphosphate in the planting furrow, Canta Galo farm, BA, 2018.

After the fertilizer has been added, it is mixed into the furrow with a three-rod subsoiler, two rods at the front, closing the furrow, and one rod at the back, in the center of the furrow, mixing the fertilizer (Fig. 5).

Figure 5 Three-rod subsoiler, Canta Galo farm, BA, 2018.

Once the fertilizer has been incorporated into the soil, the holes are marked with a 2.25 meter pole, which represents the distance between plants in the planting line, and then 500 grams of peat is added per hole, a plant material rich in organic matter, which is mixed into the soil by hand, using hoes. The holes are then dug with the help of a wooden pestle, which is pounded into the soil to the desired depth for planting the seedlings, varying between fifteen and twenty centimeters.

3.2. Breaking seed dormancy

The seeds used in Brazil to plant the Formosa variety of papaya are imported from Taiwan, since there has been no progress in the genetic improvement program in Brazil to produce national hybrids. As a result, the "Tainung 01" hybrid, which was introduced to Brazil in the early 1970s, continues to dominate the domestic market for "Formosa" group papaya trees to this day (Fig. 6).

Figure 6. hybrid papaya seeds, 'TAINUNG 1', Canta Galo farm, BA, 2018.

Seed dormancy is a natural phenomenon whereby even in conditions suitable for germination, germination does not occur, which is detrimental to nursery activities where large quantities of seeds are expected to germinate in a short space of time, allowing for the production of uniform seedlings.

Therefore, to induce seed germination, the seeds are bathed in warm water for 12 hours in order to start the germination process by means of thermal shock and the availability of free water for absorption, which will result in the seed opening and the radicle, the embryonic root that will give rise to the seedling (Fig. 7).

Figure 7. Seeds after breaking dormancy stored in a Styrofoam box with moistened paper, Canta Galo farm, 2018.

3.3. Preparing the substrate mixture for the bag and sowing the seeds

To achieve uniform and vigorous germination, in addition to using good quality seeds, it is important to provide a favorable environment for the seeds, where the humidity is balanced and the amount of water does not affect aeration, preventing the availability of oxygen.

Taking these factors into account, a mixture of substrate, soil, fertilizer and limestone is made to improve the initial development of the seedlings and improve the aggregation of the soil to the roots, forming a uniform clod that will not break when the seedlings are transplanted from the bag into the soil. This mixture is made in the proportions of two parts soil, half part substrate, half part organic peat fertilizer, two kilos of simple superphosphate mineral fertilizer and two kilos of limestone (Fig. 8).

Figure 8. Preparation of the mixture, Canta Galo farm, BA, 2018.

Once the mixture is ready, 10x18 cm plastic bags suitable for seedling production are filled by hand and the sowing process begins. This consists of sowing two seeds per bag, which have already undergone the process of breaking dormancy and have started their germination process, requiring only two or three daily waterings to obtain quality seedlings.

The plastic bags are placed in a nursery with a fifty percent shade screen, as the seedlings cannot yet withstand the sun's radiation during the hottest hours of the day, which could harm their development (Fig. 9).

Figure 9. Farm nursery, Canta Galo farm, BA, 2018.

3.4. Planting seedlings in the field and sexing

The seedlings that have been produced in the nursery, once they have reached the appropriate size to be transplanted to the field, characterized by the emission of the third leaf pair, are soaked in water to improve the aggregation of the soil particles in the bag. This process prevents the clod from breaking during transplanting and consequent breakage of the roots, which damages the quality of the seedlings when they are removed from the plastic bag for planting in the field.

During the planting of the seedlings, the bags are torn superficially with the help of a stylus, so as not to break the nearby roots, and then the clod of earth is carefully removed and placed inside the previously opened holes, so that the clod is at ground level, avoiding suffocation of the seedlings or loss of roots due to the action of rain on the exposed clod (Fig. 10).

Figure 10. Planting seedlings in the field, Canta Galo farm, BA, 2018.

Once they have been taken to the field, two small bags (with two seeds each) are placed in each planting hole, since four plants per hole are used to grow papaya at the Canta Galo farm until the flowering stage, since the hybrid seeds give rise to plants of two sexes, female plants and hermaphrodite plants, but only hermaphrodite plants are desired in cultivation, since female plants produce papaya with no commercial value. The sex of the flowers on the papaya tree (Carica papaya L.) determines the shape of the fruit. Hermaphrodite flowers form elongated fruit and female flowers form rounded fruit. However, the *fresh papaya* market only consumes fruit with an elongated shape (Fig. 11).

Figure 11. On the left, three specimens of hermaphrodite flowers of different sizes. On the right, two female flowers, characterized by the absence of stamens (male organ), Canta Galo farm, BA, 2018.

Therefore, in order to guarantee the plant stand and increase the likelihood of having at least one hermaphrodite plant per hole, four seedlings are taken together until the moment of flowering, when the female plants are identified and removed, thus forming a plantation with only hermaphrodite plants, which will produce marketable fruit (Fig. 12).

Figure 12. Management of four plants per hole until flowering, Canta Galo farm, BA, 2018.

For example: if three female plants and one hermaphrodite appear in the pit, eliminate the three females. When two hermaphrodite plants and two female plants appear in the pit, eliminate the two females and the smaller hermaphrodite. If all the females appear, remove all the plants from the hole and replant four more seedlings. And if all four hermaphrodites emerge, the three smaller ones are eliminated.

The process of eliminating unwanted plants described in the previous paragraph is called sexing. This operation is carried out approximately ninety days after planting the seedlings, when the papaya tree blooms, making it possible to identify the sex of the plant by its flower. The sexing process is a manual operation in which the worker, using a knife, cuts off the unwanted stalks close to the ground. It is a stage of great economic importance for the crop, as it ensures that only apples with commercial value are produced, guaranteeing the profitability of the crop.

3.5. Fertilizing

The papaya tree is a crop that absorbs and exports a large amount of nutrients and has continuous requirements during the first year. As it is a crop characterized by intermittent harvests, from the start of production, the plant requires its water and nutrient needs at frequent intervals, in order to allow continuity in the production of flowers and fruit (OLIVEIRA et al., 2004).

Top dressing is carried out monthly after planting, the most important being at ninety days, after the sexing stage, when the plant makes better use of the nutrients, since unwanted seedlings have been eliminated from the pit, reducing competition between plants for light and fertilizer. During the first year, until the fruit sets, the papaya tree is demanding in macro and micronutrients, which are essential for the vegetation, flowering and fruiting stages.

Potassium is responsible for providing larger fruit, with higher sugar and total soluble solids contents, and nitrogen for forming chlorophyll, the pigment responsible for photosynthesis and consequently plant growth. However, a high ratio of N/K2O nutrients in the soil should be avoided, as the plant may grow excessively, and the fruit may have a thin, soft skin and an altered taste.

Calcium is the third most required nutrient by the papaya tree, as it promotes root growth and multiplication, and its deficiency causes softening of the fruit pulp, resulting in less resistance to transportation and a shorter shelf life. Phosphorus, on the other hand, is the macronutrient required by the papaya tree in the smallest quantities. It is most important in the initial phase of root growth and is related to the fixation of the fruit on the plant. Its deficiency initially appears in the older leaves, which show yellow spots along the edges.

Another nutrient needed by the crop is magnesium, which plays an important role in the process of photosynthesis and in the absorption and translocation of phosphorus, as well as sulphur, which plays roles in the plant that determine increases in fruit production and quality (OLIVEIRA et al., 2004).

With regard to micronutrients, boron is the most important for papaya, as it is extracted in large quantities and is related to fruit quality and production. In the case of severe boron deficiency, the fruit looks lumpy and poorly formed, and in periods of drought, flower abortion can occur.

In order to determine the fertilization needs of the papaya tree, it is necessary to first carry out a chemical analysis of the soil in order to know the nutrient levels in the area.

(availability and deficiencies). When planting is established, in addition to the chemical analysis of the soil, it is important to carry out leaf analysis to confirm deficiencies or to know the nutritional status of the plants. By knowing the availability of nutrients, the fertilization recommendation can be defined more correctly.

Top dressing can be done manually or mechanized, so that the fertilizer is distributed in the planting line between plants or per hole, depending on the fertilizer to be used and its purpose (Fig. 13).

Figure 13. Fertilizer entering the area, Canta Galo farm, BA, 2018.

The nutritional needs of the papaya tree are not the same throughout its cycle, as it is a fast-growing and constant plant. The crop is fertilized at frequent intervals, giving preference to soluble fertilizer sources. Fertilizing is a practice whose success depends on the timing of the fertilizer application and the right amount and location.

Top dressing with solid fertilizers should always be done with good soil moisture. Nitrogen and potassium are fertilized monthly, while phosphorus is fertilized every two months.

The most commonly used fertilizers are potassium chloride and potassium sulphate. Fertilizers with micronutrients can be applied in the hole, as a top dressing on the soil or by foliar application. In the event of a boron deficiency, or if boron was not applied at planting, use a 0.25% boric acid solution, spraying leaves once a month.

3.6. Cleaning the streets and planting lines

In the case of papaya plantations, competition between the main crop and the weeds in the area should be avoided in order to increase productivity, since the availability of water, nutrients and space is greater in the absence of weed competition. During the initial stage, when the papaya plants are still small, weed control is carried out manually, using hoes to weed the planting lines, and close to the papaya stalk, the weeds are removed by hand, an operation known as "crowning the seedlings", in order to avoid any physical damage that may occur to the plant's stem due to the use of the hoe (Fig. 14).

Figure 14. Area after manual weeding of the planting lines, Canta Galo farm, BA, 2018.

In the streets, between the planting rows, control is carried out with a brushcutter, as the use of herbicides in the early stages of the crop causes phytotoxicity in the plants.

In crops with more developed suckers, where the plants have more lignified and resistant stems, herbicides can be used in the planting lines without phytotoxicity occurring, thus eliminating manual work with hoes and optimizing the process with the use of the tractor-sprayer combination (Fig. 15).

Figure 15. Tractor-sprayer set applying herbicide in the planting line, Canta Galo farm, BA, 2018.

The cleaning of the streets and planting lines is carried out throughout the production cycle of the papaya tree, and also has an influence on fertilization, since the cleaner the area, the better the distribution of the fertilizer and its absorption efficiency by the main crop, since there will be no competition with invasive plants, thus increasing the availability of nutrients (Fig. 16).

Figure 16. Area after herbicide application along the planting lines, Canta Galo farm, BA, 2018.

3.7. Enjoy

The characteristic of the papaya crop at flowering time is the production of imperfect flowers in the midst of perfect flowers, known as carpeloid flowers, which in the future will give rise to carpeloid fruits, which have no commercial value due to their poor shape and undesirable appearance (Fig. 17).

Figure 17. Carpeloid flower of the papaya tree, Canta Galo farm, BA, 2018.

The appearance of these deformed flowers is related to genetic factors, which are regulated by environmental factors such as higher altitudes, lower minimum temperatures, particularly temperature changes, excess nitrogen and soil moisture.

These deformed flowers continue their development until fruiting, giving rise to distorted fruit which, throughout their development, will compete for nutrients and especially space with the healthy fruit, which is not interesting in a commercial papaya crop, as these distorted fruits have no market value and are neither harvested nor sold by the producer (Fig. 18).

Figure 18. Still small carpeloid fruit, Canta Galo farm, BA, 2018.

To date, there is no alternative for controlling the appearance of these carpeloid flowers, as it is related to climate and genetic factors. Therefore, the regular practice of de-stemming has been adopted, which consists of manually removing the carpeloid fruits in order to thin out the bunch and improve the development of the healthy fruits, which will have more space to develop and consequently more nutrients available compared to the densely packed bunch, where no de-stemming has been carried out. The fruit is removed when it is still small and green.

In practice, de-stemming is done with the help of platforms, which are fitted to harvesting carts and pulled by a tractor. The purpose of these platforms is to help the worker reach the bunch of papaya when it is already developed and over two meters high (Fig. 19).

The workers, one on each side, climb onto the platforms and enjoy the bunches as the tractor driver moves the tractor-trailer, passing by each planting street and removing the carpeloid fruits with their hands, which are then thrown to the ground and rot there, mixing with the soil. In smaller plantations with younger plants, the use of platforms and tractor-trailers is not necessary, as the bunches are low and the fruit can be reached by walking alongside the plants.

Once it has been cleared, the operation is only repeated when the plant produces new carpeloid flowers, which give rise to distorted fruit, thus requiring the staff to clear the area.

3.8. Weeding / Defoliation

In addition to the Enjoyment practice, seen in the previous topic, there are other practices that aim to improve the development of the plant and lead the crop to higher yields, among them weeding and defoliation.

De-sprouting consists of removing the lateral shoots emitted by the main stem of the papaya tree. This procedure avoids reducing the growth of the plants and makes it possible to grow them without branches, which are not interesting for cultivation as they compete for nutrients and water, as well as becoming a focus for pests such as the spider mite and the whitefly (Fig. 20).

Figure 20. Papaya plants with lateral shoots on the stem (A). Papaya plants after the de-sprouting operation (B), Canta Galo farm, BA, 2018.

Weeding is done manually, with the help of a wooden stake in the shape of a ruler, during which the workers walk along the planting line, and by scraping the main stem, they remove the lateral shoots plant by plant. This practice is carried out throughout the plant's cycle whenever necessary.

Defoliation, which is the operation of cutting off the leaves, is also considered a cultural treatment. It is carried out as the old leaves senesce, and its purpose is to aerate the plants, eliminate sources of inoculum and improve the efficiency of spraying, since old leaves can overlap the younger ones at the time of application, thus reducing the area sprayed and reducing the efficiency of the product. Therefore, leaves that have fallen or have broken petioles and chlorotic leaves are removed using a knife, where the employee walks along the planting line, defoliating plant by plant manually (Fig. 21).

Figure 21. Plant with old, chlorotic leaves (A). Clean plant after defoliation (B), Canta Galo farm, BA, 2018.

Another advantage of defoliation on smaller plants is to prevent leaves that have senesced and fallen off from becoming a source of fungal inoculum for healthy leaves and consequently for the fruit, because on smaller plants, the fallen leaves are closer to the ground, a region of greater humidity, which is conducive to the development of fungi that attack the papaya crop, such as Variola, popularly known as Pinta-Preta. After rainy periods, when soil aeration is reduced, the papaya tree shows chlorosis of the old leaves, which is the ideal time to carry out defoliation.

3.9. Virus monitoring / Roguing

The main group of diseases that affect the papaya tree are viruses, which cause major losses in the plant stand, which can lead to the total destruction of the affected plantations, as well as causing constant changes in the production areas and pushing them further and further away from the consumer market. Viruses cause serious damage to the cultivation of *papaya* in Brazil, and the main viruses infecting *papaya* in the region where the internship was carried out, considering the Canta Galo farm, located in the state of Bahia, are: *Papaya ringspot* virus (PRSV) and Meleira *virus,* which is currently being characterized.

Ring spot virus, or mosaic virus, is the disease of greatest economic importance and the one with the widest geographical distribution. Its occurrence in Brazil was first reported in the state of São Paulo in 1969 and then in Ceará in 1973, with occurrences in all regions of Brazil. A production field can have 100% of the plants infected between four and seven months after planting, if no form of control is used (LIMA; LIMA, 2002). The disease manifests itself in the form of mosaic symptoms, leaf distortion, oily spots on the petioles and oily rings on the fruit, which are the main symptoms of the disease (Fig. 22). The diseased plants show whitening of the veins and the youngest leaves curl downwards one to two weeks after inoculation. Over the course of a few weeks, the leaves become mottled and distorted, with the lobes greatly reduced in size (Fig. 23).

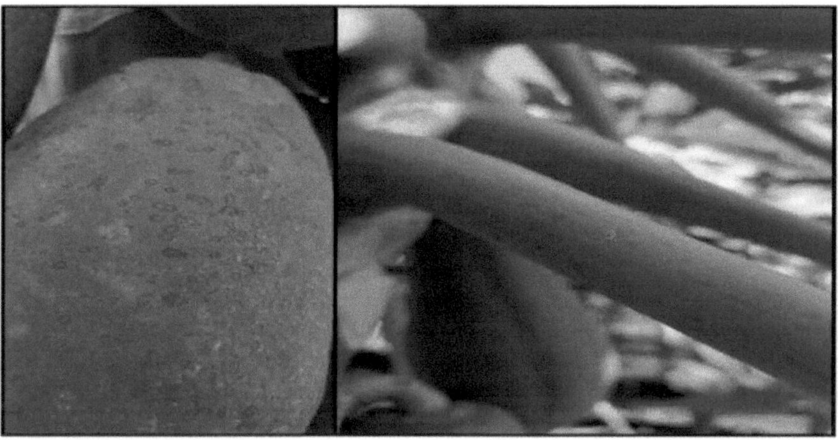

Figure 22. Symptoms of the virus on the fruit and petiole of a papaya plant, Canta Galo farm, BA, 2018.

Figure 23. Virus symptom on apex leaves, Canta Galo farm, BA, 2018.

The virus is not transmitted by seed, but by different species of aphids. Although some of these aphids do not colonize papaya plants, the bite of a test is enough for the process of acquiring and inoculating the virus to take place.

In 1989, the late blight virus was first reported in the state of Bahia and then in Espirito Santo. Since then it has spread rapidly, causing damage to entire plantations, becoming a limiting factor for breast production (LIMA; LIMA, 2002).

The disease is characterized by the exudation of latex from the fruit, which oxidizes, resulting in a "blurred" and "honeyed" appearance, giving the disease its name (Fig. 24). These symptoms also appear on the petioles and margins of new leaves, before fruiting, which become necrotic after the latex exudes. The fruits are poorly formed, with light green zoned spots, depreciating their commercial value. The latex of the fruits of plants with late blight is watery, does not coagulate and therefore runs off easily (LIMA; LIMA, 2002).

Figure 24. Plant with Meleira virus symptoms, Canta Galo farm, BA, 2018.

So far it has not been possible to conventionally transmit the blight virus mechanically to papaya, but the virus has been transmitted to healthy plants via wounds with blades soaked in latex from infected plants. It is not yet known how the virus is transmitted in the field. Various strategies have been adopted to control papaya viruses, including those described below:

Avoid intercropping with cucurbits, as the virus is capable of infecting them, in order to avoid the possible establishment of virus and aphid sources, due to the aphids' preference for cucurbits. It is advisable to keep orchards clean, especially if they are free of native species of cucurbits, in order to avoid the formation of aphid colonies and possible sources of the virus. Disinfect tools, especially knives and pruning shears, with a 1:10 solution of sodium hypochlorite (bleach) / water (LIMA; LIMA, 2002).

And finally, Roguing, an indisputable practice for controlling viruses, is the most important of all the practices described above. Roguing is the most efficient method of preventing the spread of viruses in the field, which consists of eradicating contaminated plants, as there is no cure for plants once they have been infected by the virus, whether it's blotch or ring spot.

Roguing is practiced on the Canta Galo farm on a daily basis, where employees known as viroseiros are responsible for diagnosing and eliminating contaminated plants. The virologists walk along the planting lines and diagnose the plant visually for ringspot virus. They are trained employees who have practice in identifying diseased plants. For the ringspot virus, when a plant is identified with symptoms of oxidized exudate on the fruit, confirmation is made with the help of a pointed stick, which they use to pierce the fruit and analyze the viscosity of the exudate released; if it is watery, a characteristic symptom of ringspot, the plant is eliminated.

When the virologists identify a contaminated plant, there is a Roguing procedure in which they cut the papaya stalk off close to the ground with a machete and then pick up the fallen plant and cut off its leaves as well, thus accelerating the wilting of the plant and the consequent death of the tissues. Once the plant has been cut, it is collected and lined up in a planting line to prevent tractors from driving over the contaminated tissues, which could lead to the virus spreading to other areas (Fig. 25).

Once the area has been rogued, the viruses return to the area within a period of seven days, which is enough time for the viruses to manifest themselves in previously infected plants that didn't show any symptoms, and this is how the cycle of eradicating plants with viruses, known as roguing, takes place.

Figure 25. Infected plant after roguing, Canta Galo farm, BA, 2018.

3.10. Pest monitoring and control

In papaya growing, there are several pests that attack the plants and cause injury, but not all of them mentioned in the literature occurred on the Canta Galo farm during the internship period. The most common insects to occur on the farm and their respective forms of control are described below.

The most common pests were: whitefly (Bemisia-tabaci), spider mite (Tetranychus urticae), mealybug (Aonidiella comperei) and green leafhopper (Empoasca sp). The pest monitoring carried out on the farm aims to identify the level of insect infestation so that decisions can then be made about the application of insecticides and their dosages. This monitoring is carried out by an employee who, while walking in a zigzag pattern through the area, examines ten plants at ten different points in the chosen plot, writing down the level of pest infestation on the field record. Papaya is a crop that is very sensitive to the phytotoxicity of the products used in the chemical control of pests, so for effective control, it is necessary to observe the type of pesticide, the pest to be controlled, the time of application and the incidence of natural enemies in the area.

The whitefly, a common insect in papaya plantations, is not recognized as an important pest in papaya cultivation, and the level of control of this insect is not measured, since its damage to the crop is minimal and indirect. The flies have a habit of inhabiting the tops of papaya trees, specifically the underside of the new leaves. When fumagina develops on the leaves due to fly attacks, the insect is chemically controlled to avoid losses due to a lack of photosynthesis in the plant, as the fumagina fungus prevents the leaf from breathing in the area where it develops.

To control flies, apply an insecticide registered for the crop to the infested area, using a tractor-turbo sprayer, and adjust the flow rate according to the recommendations on the product's leaflet.

The spider mite is one of the most important insects for the papaya tree, as it causes serious injury and damage to the crop. It attacks the underside of older leaves and forms colonies between the veins closest to the petiole. They are insects that, when they feed, destroy the cells of the leaf tissue, turning the leaves yellow on the opposite side of the attack, later, as the wound develops, these areas become necrotic and the leaves become brittle. In severe infestations, the leaves fall off and the plant produces lateral shoots, which is undesirable, as well as leaving the fruit exposed to the sun due to the absence of leaves, also causing indirect damage.

The level of control for the spider mite is low, and it is necessary to carry out chemical control as soon as colonies develop on the underside of the leaves, as they are insects that reproduce and spread quickly.

Mealybugs can be found in large colonies on the stem of the papaya tree or on the fruit. They are sap-sucking insects and also cause damage to the papaya tree, in addition to depreciating the infested fruit, losing its commercial value, as they are insects that adhere to the fruit and are difficult to remove. Chemical control is carried out as soon as their presence is detected, and consists of scraping the stem to better expose the mealybugs and then spraying with 1% emulsifiable oils or sulphocalcic syrup.

Finally, another insect of great economic importance to the papaya tree is the green leafhopper, a small, sap-sucking insect that moves laterally, making it easy to identify. The adults and nymphs are found on the underside of older leaves sucking the sap. Continued sucking by the leafhoppers causes yellowish spots at the site of the bite. Chemical control is carried out when leafhoppers are found on 10% of the leaves examined, and an insecticide registered with the Ministry of Agriculture, Livestock and Supply (MAPA) for papaya is applied, at the dosage and rate found on the product leaflet (Fig. 26).

Figure 26. Application of insecticide with turbo sprayer, Canta Galo farm, BA, 2018.

It is worth emphasizing that monitoring pests is of great importance for their successful control, as it is not always necessary to spray the area with insecticides, as there are natural enemies of these pests that are also harmed by spraying, which is why control levels should be respected and pest sampling should be carried out periodically.

3.11. Disease monitoring and control

As with the previous topic on pests, there are many diseases that attack the papaya crop, but we will only describe the diseases that occurred on the Canta Galo farm during the internship period, including: Pinta-preta or variola, Anthracnose (post-harvest) and Podridao-de-Phytophthora or gummosis.

Diseases represent a major obstacle to production, as they compromise the quality and quantity of fruit produced. Chemical control is an important tool to be used to reduce the incidence of diseases, but, as with pests, it should only be used when the diseases have reached a certain level of incidence or damage, defined as the control level. As with insecticides, the fungicides used must be registered, taking into account their efficiency, maximum permitted residue limit and grace period for consumption after spraying.

Black spot or variola is the most common disease of the papaya tree, and symptoms occur on both the leaves and the fruit. Its initial appearance is on the leaves, on the upper page there are round necrotic spots surrounded by a yellow halo. On the lower page of the leaves, in the areas corresponding to the chlorotic spots, you can see the powdery growth of the fungus, similar to black dots (Fig. 27).

Figure 27. Black dot symptoms on old papaya leaves, Canta Galo farm, BA, 2018.

Young leaves generally don't show symptoms. If left unchecked, the fungus spreads to the fruit, where it causes lesions of varying sizes, depreciating it commercially. These lesions are usually epidermal, but in severe cases they can reach the flesh of the fruit. The etiological agent of black spot is the fungus *Asperisporium caricae*. This fungal disease can occur in all months of the year, with greater intensity in the rainy months and with strong winds, as these conditions favor the development of lesions and the dispersal of spores from old leaves, which are considered sources of inoculum. When monitoring the plants, look at the oldest leaves, where the lesion first occurs, spraying with fungicides should begin as soon as the first symptoms are observed.

The application intervals depend on the climatic conditions of the region, but at the Canta Galo farm they are carried out every two weeks, preventively and curatively, always combining two chemical groups of fungicides, such as strobirulins and triazoles. It is important to alternate the active ingredient of the fungicides to prevent the pathogen from developing resistance

Anthracnose is one of the main post-harvest diseases of papaya and occurs in all producing regions. Fruits infected by the fungus *Colletotrichum gloeosporioides* become unfit for sale and the symptoms develop during post-harvest, which is the appearance of dark spots on the peel, cylindrical or elliptical in shape, forming a gelatinous mass which, if removed from the peel, forms a hole in the fruit (Fig. 28).

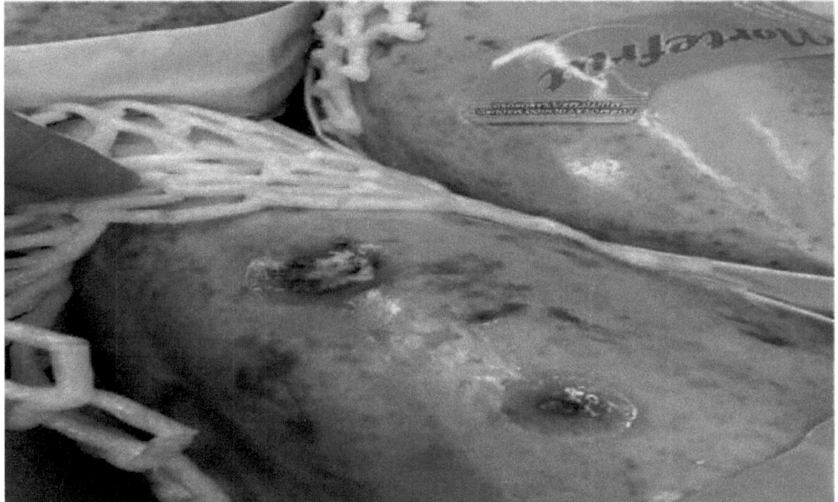

Figure 28. Anthracnose symptoms on post-harvest fruit, Packing house, BA, 2018.

Infection usually begins in the field, where the pathogen remains unnoticed on the fruit until ripening, during which time it develops and symptoms appear. As with black spot, high relative humidity favors the development and spread of the fungus, which needs free water for the germination of conidia, the fungus's form of reproduction. The severity of the disease is lower in cold, dry months.

In order to prevent the fungus from manifesting itself on the fruit in the post-harvest period, measures should be adopted during the production and processing phases, such as careful handling, asepsis of packaging and the workplace, acclimatization of the harvested fruit if possible, and chemical control in the field. Fungicides are also applied every fortnight, just like for black spot, in order to prevent the spread of the pathogen and keep the level of infestation low. Fungicides from the dithiocarbamate group and chlorothalonil are used, sprayed on the fruit pre-harvest to control anthracnose, as well as fungicides used post-harvest for value-added fruit that is sent to the packing house. Spraying begins at the time of flowering and can vary according to the time of year.

Pytophtora rot, also known as gummosis, occurs mainly in rainy periods, moist and poorly aerated soils. Its damage can be seen in the neck, fruit and roots, and when its incidence is severe in the neck, it causes the death of the plant. It can cause serious damage to the crop when its incidence occurs in the fruit, which is characteristic of rainy periods, when rotting is observed in the tissues, especially of ripe or almost ripe fruit. This rot covers the fruit tissues with white mycelia, forming a foam-like symptom, which in the Bahia region is called "Santa Claus" (Fig. 29). The pathogen of the fungus is *Phytophtora palmivora*, and to avoid its incidence it is not recommended to plant in very clayey, poorly drained soils or regions with a lot of rain. When the weather conditions are favorable, preventive spraying with copper-based fungicides is recommended, and in cases where the fungus occurs, apply "Phytogard" to the neck of the plant, the leaves and the fruit in a curative manner.

Figure 29. Symptoms of Pythophora rot on the stem on the left and on fruit fallen to the ground on the right (Santa Claus), Canta Galo farm, BA, 2018.

3.12. Preparing the mixture for spraying

The preparation of the mixture to be sprayed is of great importance to the success of the application, because if products are mixed in the wrong order, efficiency is not guaranteed. On the Canta Galo farm, the mixture is prepared as follows: the employee adds water up to half the tank's capacity, then turns on the mixture mixer, which is located inside the spray tank. With the mixer on, the adjuvant product is added to the water, which is designed to reduce the pH of the water to improve the efficiency of the product used, as well as having a buffer function, which keeps the pH of the mixture low during application. Once the pH reducer has been added, the product is added, be it an insecticide, fungicide or herbicide, always measured in a suitable dosing device with volume markings to guarantee the recommended dose, and finally the tank is filled with water and the mixture is ready to go into the field (Fig. 30).

Figure 30. Spray mixture preparation, Canta Galo farm, BA, 2018.

The employee who prepares the syrup always uses personal protective equipment (PPE) for their safety, since they come into contact with highly toxic chemicals. The PPE used is gloves, boots, an apron and full clothing.

3.13. Harvest / Domestic market / Exports

One of the most important operations for guaranteeing the quality of papaya is harvesting. Papaya is harvested by hand, as the fruit is highly perishable and requires careful removal and handling. The way the papaya is harvested depends on the destination of the fruit: it can be for the domestic market or for the packing house, which exports it and also sells it to restricted markets that are more demanding in terms of quality. Papaya is harvested on a weekly basis, since the availability of fruit is constant and the time of harvest is defined by the stage of ripeness of the fruit, which is characterized by the formation of yellow stripes on the skin (Fig. 31).

Figure 31. Fruits with yellow stripes that have reached harvest point, Canta Gaio farm, BA, 2018.

If the harvested apples are destined for the domestic market and transported in bulk, the harvest is quicker and less careful, where the workers remove the fruit from the bunch with their hands and place it in the trailer which is hooked up to the tractor. The tractor-trolley combination moves between the planting rows and the workers walk alongside to remove the fruit that has reached the ripening stage. This cart has a capacity of approximately two thousand kilos.

When harvesting takes place in older areas, where the fruit is over two meters high due to the continuous growth of the plant, the workers harvest on top of harvesting platforms which are attached to the carts. These platforms lift the worker up so that he can easily reach the bunch of fruit. They are fixed, height-adjustable platforms which are easy to assemble and offer safety to the worker (Fig. 32).

Figure 32. Harvesting platforms attached to the trailer, Canta Galo farm, BA, 2018.

There are always two employees harvesting, as the tractor moves in the middle of the planting rows, allowing each employee to harvest on each side.

In this case, when the platform is used, two more workers are needed inside the truck to eliminate the height at which the harvested fruit can fall. These workers receive the fruit one by one to prevent it from hitting the floor of the truck, minimizing impact damage.

Once the carts are full, the workers remove the harvesting platforms and wait in the field, while the tractor driver takes the harvested papaya to the shed and returns with another empty cart to continue harvesting (Fig. 33).

Figure 33. Cart full of papaya being taken to the shed, Canta Galo farm, BA, 2018.

The fruit that arrives at the warehouse is loaded onto the truck by another group of employees called packers. The packers wrap the fruit in newspaper to protect it during transport, and then load it, either in bulk or in wooden crates that can hold approximately fifteen kilos (Fig. 34).

Figure 34. Bulk mammon load in a truck (A), Canta Galo farm, BA, 2018.

Figure 34. Load of papaya in wooden crates on a baù truck, Canta Gaio farm, BA, 2018.

When the fruit to be harvested is destined for processing in the Packing house, with the aim of adding value to the fruit, the harvest is more rigorous in terms of care and standardization. Fruit destined for processing in the packing house can be exported or sold to large markets, which demand better physical quality from the fruit. Therefore, at the time of harvesting, when the employees remove the papaya, it is already placed in an individual bubble wrap bag to avoid any damage to the fruit due to contact with each other during transportation to the packing house. Once the fruit has been protected in the bubble bag, it is placed in container boxes (Fig. 35). These container boxes can hold approximately fifteen kilos, preventing the fruit from weighing too much on top of each other, unlike harvesting for the domestic market, where the fruit is piled up on the truck.

Figure 35. Breast harvest for export, Canta Galo farm, BA, 2018.

These container boxes can hold approximately fifteen kilos, preventing excessive weight of fruit on top of each other, unlike harvesting for the domestic market, where the fruit is piled up on the truck.

Once the container boxes have been loaded onto the truck, the tractor driver takes them to the shed, where they will be transferred to a truck, which will transport the fruit to the packing house.

3.14. Post-harvest/ Packing house

Post-harvest is the stage where value is added to the fruit, which is washed, waxed and packed in paper boxes to reach more demanding consumer markets that pay more for better quality fruit. The packing house is located outside the Canta Galo farm, but not too far away so as not to increase processing costs. There, the fruit is unloaded from the boxes and processed in a suitable machine, eliminating contact between people and the fruit until it is packed. First, the fruit is dumped into a washing tank, where it comes into contact with a mixture of water and Ecotarget, a disinfectant product added to the washing water at a concentration of 200 ml per 400 liters of water (Fig. 36).

Figure 36. Fruit in the washing tank at the start of the processing process, Packing house, BA, 2018.

They then go on to a continuous conveyor belt where the fruit is scrubbed and mechanically washed to remove dirt and improve the visual appearance of the fruit (Fig. 37).

Figure 37. Scrubs cleaning the fruit, Packing house, BA, 2018.

After washing, the fruit goes onto the conveyor belt where it is mechanically sprayed with fungicide to prevent anthracnose from appearing on the fruit during ripening. The fungicide used is Magnate 500EC, at a concentration of 50 ml per 100 liters of water (Fig. 38).

Figure 38. Fruit on the conveyor belt heading for the spraying process, Packing house, BA, 2018.

After the protective fungicide has been applied to the fruit, it goes through a cold ventilation to dry the product and then, on the same conveyor belt, it is sprayed again with a mixture of water and food wax in the proportion of 24 liters of wax to 400 liters of water, in order to improve the visual aspect of the fruit and, finally, it goes through drying again with a cold ventilation to be packaged (Fig. 39).

Figure 39. Spraying food wax on fruit, Packing house, BA, 2018.

Once the fruit has left the conveyor belt, already processed and dried, the employees wrap the fruit in a protective net made of a foamy material called expanded low-density polyethylene (LDPE) and also place stickers with the company's logo on each of the fruits, which will be placed in the cardboard boxes destined for the market (Fig. 40).

Figure 40. Cardboard boxes with the fruit already processed, ready for sale, Packing house, BA, 2018.

The employees are also responsible for disposing of fruit that is unfit for consumption or has mechanical damage to the skin, since the processing machine doesn't have this function.

After being properly packed in cardboard boxes, the boxes are taken to the cold room, which keeps the temperature below 15 degrees Celsius, the ideal temperature for the fruit to stop its metabolism and extend its shelf life, as low temperatures slow down the ripening of the papaya (Fig. 41). This concludes the processing process, with the fruit ready to be sold.

Figure 41. Boxes with fruit stored in cold room, Packing house, BA, 2018.

4. FINAL CONSIDERATIONS

1. One of the main challenges for the papaya crop is to control the viruses and maintain the plant stand during the production cycle, as the viruses spread rapidly in the crop and there is no curative method, only the eradication of the plant once it has been contaminated.

2. The main bottleneck in the papaya production stage is the harvest, as the fruit sold differs in size and stage of ripeness according to the customer, making it a difficult stage to adapt to in order to maintain a constant supply of fruit.

3. The main tool for success in an enterprise like the Canta Galo farm is good management, where the administrative staff and employees work together to manage and administer the business.

4. The demand for better quality fruit has increased and the future scenario indicates that this fruit will be in demand in the big markets, where those that don't fit in will find it difficult to sell their produce.

5. The compulsory curricular internship for obtaining the title of agricultural engineer, carried out at the Canta Galo farm, supervised by agricultural engineer Marcos Souza Leal, was very important for applying basic theoretical knowledge and developing new ideas, as well as their application in the day-to-day running of the farm.

5. BIBLIOGRAPHICAL REFERENCES

BADILLO, V.M. Caricaceae: second scheme. **Review Facultad de Agronomia,** Maracay, v.43, p.1-111, 1993.

BADILLO, V.M. **Monografia de la familia Caricaceae.** Maracay: Nuestra América, 1971. 221p.

CARDOSO, Mauricio Mendes; PACHECO, Dilermando Dourado. Edaphoclimatic requirements: Final considerations. **Informe Agropecuârio**: Cultivo do Mamoeiro, Belo Horizonte, v. 34, n. 275, p.27-27, jul./ago. 2013.

CARDOSO, Mauricio Mendes; PACHECO, Dilermando Dourado. Edaphoclimatic requirements: Introduction. **Informe Agropecuârio**: Cultivo do Mamoeiro, Belo Horizonte, v. 34, n. 275, p.25, jul./ago. 2013,

COSTA, Adelaide de Fâtima Santana da et al. Botany, breeding and varieties: Taxonomic aspects. **Informe Agropecuârio**: Cultivo do Mamoeiro, Belo Horizonte, v. 34, n. 275, p.14-15, jul./ago. 2013.

FRANCO, Marcelo Lana. Quality papaya in new production areas. **Informe Agropecuârio**: Cultivo do Mamoeiro, Belo Horizonte, v. 34, n. 275, p.3-4, jul./ago. 2013.

LIMA, Roberto C. de Araùjo; LIMA, J. Albérsio de Araùjo. The war on papaya viruses. **Cultivar Hortaliças e Frutas,** Cearà, v. 14, p.3-12, jun. 2002. Available at: <http://www.grupocultivar.com.br/artigos/guerra-as-viroses-do-mamao>. Accessed on: March 12, 2018.

LYRA, G.B. Estimation of the optimum economic levels of irrigation and nitrogen fertilization in papaya trees (Carica papaya L.) cultivar Golden and hybrid UENF Caliman 01. 2007. 160f. Thesis (Doctorate in Plant Production) - Universidade Estadual do Norte Fluminense "Darcy Ribeiro", Rio de Janeiro, 2007.

OLIVEIRA, A.M.G et al. **Breast for export:** technical aspects of production. Brasilia: EMBRAPA-SPI, 1994. 52p. (FRUPEX. Technical Publications, 9).

OLIVEIRA, Arlene Maria Gomes et al. Nutriçâo, Calagem e Aduçâo do Mamoeiro Irrigado. **Circular Tècnica,** Cruz das Almas, Ba, v. 69, p.2-8, Aug. 2004. Available at: <https://www.agrolink.com.br/downloads/adubaçâo nutriçâo e calagem do mamoeiro embrapa.pdf>. Accessed on: March 12, 2018.

SANCHES, N.F. Integrated papaya production. In: MATOS, A.P. de (Ed.). **Integrated production of tropical fruit.** Cruz das Almas: Embrapa Cassava and Fruit Growing, 2012. P.186-287.

SIQUEIRA, D.L. de; BOTREL, N. Climate and soil for growing papaya. **Informe Agropecuârio**. Mamâo, Belo Horizonte, v.12, n.134, p.8-9, feb. 1986.

46

Printed by Books on Demand GmbH, Norderstedt / Germany